# Who Has More?

by April Barth

We pick pumpkins in the fall. The girl has 5 big pumpkins.

The boy has 9 little pumpkins.
Who has more pumpkins?

5  6  7  8  9  10  3

We pick apples in the fall, too. The boy picks 12 green apples.

**12**

10     11     12     13     14

The girl picks 17 red apples.
Who has more apples?

15    16    17    18    19    20

We find red leaves. We find yellow leaves, too.

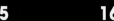

The girl finds 22 yellow leaves.
The boy finds 18 red leaves.
Who has more leaves?

The squirrels find acorns in the fall. The black squirrel finds 36 acorns.

30    31    32    33    34

The brown squirrel
finds 39 acorns. Who
has more acorns?

35    36    37    38    39    40

We play outside before dinner.
I have 40 minutes to play.

40    41    42    43    44

My sister has 45 minutes to play. Who has more time?

My friend puts 56
more seeds in
this bird feeder.

50    51    52    53    54

I have 53 more seeds for the birds. Who has more seeds?

55    56    57    58    59    60

We count our steps
in the corn maze.

65    66    67    68    69

I take 71 steps. My dad takes 67 steps. Who takes more steps?

70    71    72    73    74    75

We do many fun things
in the fall. We can count.
We can tell who has more.